Foreword

CIRIA's research programme on Trenchless and Minimum Excavation (TME) Technology is intended to provide technical guidance for all construction professionals.

The implementation from 1 January 1993 of the New Roads and Street Works Act 1991 has encouraged the development of TME techniques where there is a benefit in using these methods.

This report was prepared under contract to CIRIA by Mr J F Grimes and Mr P Martin of Binnie & Partners (now Binnie Black & Veatch), one of CIRIA's Core Programme sponsors. The research contract was guided by a steering group with the following members:

Mr K N Montague (Chairman)	Brian Colquhoun & Partners
Mr J Crossley	Yorkshire Water Engineering
Mr M DaRios and	
Mr S Byrne	BT World Wide Network
Mr M G Glynn	NWW Engineering Ltd (to December 1992)
	A E Yates Ltd (from January 1993)
Mr J E James	International Society for Trenchless Technology
Dr B M New and	
Mr G Crabb	Transport Research Laboratory
Mr J A Olver	L G Mouchel & Partners Ltd
Mr R B Rosbrook	Southern Water Services Ltd
Mr C Salts	NWW Engineering Ltd (from January 1993)
Mr P T Simpson and	
Mr G Bateman	Charles Haswell & Partners Ltd
Mr C E Tregoing	WRc

CIRIA's research manager for this project was Mr R Freer.

The project was funded by the following organisations:

Northumbrian Water Services Ltd
North West Water Ltd
Thames Water plc
Welsh Water plc
Southern Water Services Ltd
Wessex Water Services Ltd
Yorkshire Water Services Ltd
British Telecom
CIRIA's Core Programme sponsors.

CIRIA and Binnie & Partners (now Binnie Black & Veatch) are grateful for the help given in the development and completion of this project by the funders, by the members of the Steering Group and by the many individuals and organisations who were consulted.

Contents

List of Figures

List of Boxes

List of Tables

Glossary

This Glossary describes the meanings attached to terms commonly used in the guidance notes and is mainly based on the Glossary of Terms contained in the *Introduction to Trenchless Technology, Second International Edition*, dated 1992, published by the International Society for Trenchless Technology, to whom acknowledgement is gratefully made.

Auger boring	Technique for conduit installation using a non-steerable rotating cutting head attached to auger flights within a casing.
Auger TBM	Type of microtunnelling machine in which the excavated soil is transferred to the drive shaft through auger flights passing through the product pipe.
Directional drilling	Technique for conduit installation involving drilling in a shallow arc using a guided steerable drilling head, generally for long spans.
Drive shaft	Excavation from which trenchless technology equipment can be launched for the installation of conduits.
Earth pressure balance machine	TBM that applies mechanical pressure to the material to be excavated to provide temporary support to the surrounding ground.
Guided drilling	Technique for conduit installation using an excavation or soil displacement guided drilling head whose alignment is monitored by a hand-held detector on the surface above the head.
Impact moling	Technique for conduit installation using a percussive soil displacement tool to form a bore.
Jacking pipes	Pipes installed using pipe jacking techniques to form a conduit.
Microtunnelling	Technique for conduit installation using a steerable, remote-controlled tunnel boring machine by pipe jacking, the excavated material being removed either by mechanical auger or as slurry.
Narrow trenching	A technique for conduit installation by excavation of a slit trench approximately 50 mm either side of the outside diameter of the service to accommodate the pipe bedding.
On-line replacement	The breaking out of an existing service and the installation of a new conduit in the same place.
Pipe bursting	A technique for conduit installation using an expanding device to break an existing pipe from within to allow a new conduit to be inserted in its place. This is also referred to as pipe splitting.
Pipe eating	Technique for conduit installation in which an existing defective pipeline is first used as a pilot and is then broken up, removed and an enlarged excavation made for a larger pipeline.

Pipe jacking	Technique of installing a line of pipes through the ground in a previously excavated bore by means of hydraulic jacks from the drive shaft. After pushing a pipe length into the ground a new pipe is positioned and the process repeated.
Pipeline system	The interconnecting pipe network for the conveyance of fluids.
Pipe ramming	Technique for conduit installation involving a casing driven through the ground by a percussive hammer.
Product pipe	Permanent pipe used in a conduit installation.
Reception shaft (exit shaft)	Excavation into which trenchless technology equipment is driven and recovered following conduit installation.
Rehabilitation	Maintaining or upgrading performance of existing pipeline systems.
Reinforcement	The provision of an additional service, which, in conjunction with an existing service, increases the overall capacity.
Renovation	Methods by which the performance of pipeline systems is improved by incorporating the original fabric as either temporary support for a new system or an integral component of a new system.
Replacement	Methods by which a new pipeline is constructed replacing the original fabric on the same line.
Rod pushing	Technique for conduit installation using steerable piercing head, which is pushed through the ground to form a bore.
Sliplining	Insertion of a new pipe, usually of polyethylene (PE), within an existing defective pipe. Developments provide for the new pipe to be temporarily reduced in diameter before insertion, and subsequently enlarged to provide a tight fit into the original pipeline.
Slurry TBM	A type of microtunnelling machine in which soil is turned to slurry and is used to counterbalance water pressure to stabilise the face, before being pumped to the surface.
Spiral lining	Technique in which a ribbed PVC strip lining is spirally wound by machine into an existing pipeline.
Spray-on lining	Technique for applying a lining or coating by a rotating head, which is winched through the pipeline.
Thrust boring	Technique for conduit installation involving a casing driven through the ground by hydraulic rams.
Trenching (open cut)	Traditional method for conduit installation involving the excavation of a trench, laying the conduit, refilling and reinstating the surface.
Trenchless technology	Techniques for conduit installation, replacement or renovation that minimise excavation from the surface.
Tunnel boring machine (TBM)	Device used to excavate a bore by mechanical means for the purpose of conduit installation.

Abbreviations

CCTV	closed-circuit television
DI	ductile iron
EA	Environment Agency
GIS	Geographical Information Systems
GRP	glass-reinforced plastic
HAUC	Highways and Utility Committees
MDPE	medium-density polyethylene
ME	minimum excavation
OS	Ordnance Survey
PE	polyethylene
PVC	polyvinyl chloride
TBM	tunnel boring machine
TME	trenchless and minimum excavation
uPVC	unplasticised polyvinyl chloride

1 Introduction

1.1 DEFINITION

Trenchless and minimum excavation (TME) technology covers a range of techniques for installing, replacing or renovating underground service conduits while minimising or eliminating excavation from the surface. In this context trenchless technology generally refers to techniques involving discrete working areas whereas minimum excavation refers to continuous narrow trenching.

1.2 BACKGROUND TO THE STUDY

These guidance notes are concerned with TME techniques for services which are smaller than man-entry size, which for the purposes of the notes is defined as up to 1000 mm in diameter. The use of tunnelling and other systems specifically for excavations greater than 1000 mm in diameter are not considered, although many of the protocols and procedures may equally apply.

TME technology offers the opportunity to install, replace and renovate underground services with the minimum of disturbance to the surface environment and to other utilities.

The development of TME techniques was encouraged by the government-sponsored Review of the Public Utilities Street Works Act in 1985[1], commonly referred to as the Horne Report. This review recognised the growing concern at the damage caused by trenching to the road structure, including reduced design life. The social costs of trenching were also beginning to be taken into account, with traffic delays, diversions and the resultant loss of trade to shops being particularly serious, together with the environmental impact on people and their lifestyle.

The subsequent New Roads and Street Works Act 1991 seeks to reduce traffic delays and encourages clients to specify the use of faster and less disruptive construction methods.

Most urban areas in the UK were supplied with water, sewerage and gas services during the Victorian era and many of these services now urgently require reinforcement and rehabilitation. Similarly, the electricity and telecommunications networks need reinforcement and modernisation to meet increased demand.

With this potential increase in workload, particularly in urban areas, and the growing concern about the adverse environmental impact, social costs and damage to the roads, it is important that alternative construction methods to trenching are considered and adopted wherever economical and feasible. In recent years, there has been an increase in the use of TME technology, although possibly not as much as some had expected. One reason is believed to be the difficulty of recognising the potential financial benefits resulting from the speed of construction and minimal disturbance when these methods are used.

Planning is a key element in the successful implementation of all engineering projects. Until quite recently it was normal for clients to assume that underground pipes and ducts would be installed or replaced in open trenches, the actual construction method used being left largely or entirely to contractors or site crews. Now that alternatives to traditional open trenching are available, those planning and designing underground services need to consider construction methods from the outset and may have to decide at an early stage what construction method is most appropriate.

Planning and investigation requirements for the application of TME technology are not only different to those for traditional open trenching, but also vary with the trenchless techniques proposed.

1.3 STRUCTURE OF THE REPORT

These notes have been prepared to provide a series of guidelines covering planning and investigation procedures.

Section 2:   Describes the TME techniques that are available, together with the factors that affect the choice of method for installing, replacing or renovating underground services.

Section 3:   Provides guidelines on planning and investigation for the use of TME techniques.

Additional and supplementary material is provided in Appendices A to E.

Appendix A: TME techniques

Appendix B: Investigation strategies and methods

Appendix C: Design considerations

Appendix D: Costs

Appendix E: Risk assessment

Published information, from the UK and overseas, concerning the investigation, assessment and planning for the use of trenchless methods was reviewed. This literature review is available as CIRIA Project Report 63.

2 The use of trenchless and minimum excavation technology

2.1 TRENCHLESS AND MINIMUM EXCAVATION TECHNIQUES

Trenchless techniques can be used for the installation of new services, on-line replacement of existing services and renovation of existing services. Minimum excavation techniques can be used for the installation of new services only.

The various methods available are listed in Box 1, and are also discussed in Appendix A. Greater detail can be found in the *Introduction to Trenchless Technology*[2] and CIRIA Technical Note 127 *Trenchless construction for underground services*[3].

Box 1 *TME techniques*

New Installation

- Microtunnelling

- Auger boring

- Impact moling

- Rod pushing

- Pipe ramming

- Thrust boring

- Horizontal directional drilling

- Guided drilling

- Narrow trenching

On-line replacement

- Pipe bursting

- Pipe eating

Renovation

- Sliplining

- Modified sliplining

- Lining formed in place

- Spray-on lining

- Localised repairs

- Chemical stabilisation

2.2 FACTORS AFFECTING CHOICE OF METHOD

Planners and designers consider a number of factors when deciding on the most appropriate solution for a project involving the installation, replacement or renovation of underground services. These include some or all of the following:

- size of service
- length
- depth
- location
- topography
- ground (soil) conditions
- site conditions
- cost
- condition of existing service
- presence of other services and apparatus
- physical obstacles (e.g. rivers, railways, buildings)
- traffic disruption (including pedestrians)
- disruption to third parties
- installation, replacement or renovation techniques
- experience of techniques
- availability of equipment
- temporary working areas
- time of occupation
- time of installation, replacement or renovation
- access
- safety
- risk of failure
- availability of resources (power, water, drainage)
- maintenance of existing service
- reinstatement requirements
- land use
- environmental factors
- settlement.

Faced with the above list of considerations, planners and designers are called upon to produce appropriate solutions that meet a number of criteria, which are sometimes conflicting:

- the need to install, replace or renovate an underground service at least cost
- the need to minimise damage to the highway
- the need to minimise disruption to traffic and pedestrians
- the need to minimise potential damage to other existing underground services
- the need to minimise any adverse impact on the environment.

Following the privatisation of most of the utility companies it is also reasonable to assume that an additional criterion is that of maintaining company or corporate image by taking account of customers' views.

Until the advent of trenchless techniques, excavation by traditional open-cut techniques was the only method available for installation and replacement of underground services. Trenchless techniques, which are being rapidly developed and advanced, now offer alternative solutions for most installation, replacement and renovation projects.

Trenchless techniques offer many potential advantages over traditional open-cut techniques, including:

- installation of services at greater depths than would normally be considered cost-effective for trenching
- installation of services in poor ground conditions
- avoidance of many surface constraints
- diversions of other services minimised
- surface reinstatements minimised
- surface disruption including traffic disruption kept to a minimum
- reduced surface settlement – particularly important in sensitive areas e.g. under railways, motorways, services and adjacent to buildings
- increased speed of installation
- improved safety of operatives and the general public
- ability to follow a 'shortest route' option
- environmental disturbance minimised
- the quality of the finished pipeline can be superior, particularly in areas of poor ground, because the soil formation is not disturbed to the same extent.

The main technical disadvantage of trenchless techniques for the installation and replacement of services is the inability to make or remake side connections on-line without some excavation. For gravity drains and sewers this is not always a limitation, since new or replacement services can be constructed off-line and connected into the existing services at chambers provided at shaft or drive pit locations. Some trenchless techniques used for renovation purposes do enable lateral connections to be remade remotely from within the pipe being renovated.

Experience has shown that to gain the full benefits of using trenchless techniques it is essential to plan, investigate and design specifically for these techniques. If the design follows the conventions of traditional open-cut trenching and is subsequently changed to trenchless during the procurement stage then many of the benefits may be lost. If a change to TME techniques is to be made at a later stage in the procurement of the work, many of the factors referred to above must be reviewed.

New or replacement sewers and surface water drains and the associated manholes and chambers are often the most disruptive services and structures to be constructed in a congested urban environment. They are normally larger and need to be deeper than most other services; they are located mainly under roads; and they must follow fixed gradients irrespective of ground levels. Microtunnelling is therefore an attractive alternative to open-cut techniques because disruption to traffic and other services is restricted to drive and reception shaft locations. The positions of these shafts are determined in advance by the designer, and locations can be selected to further minimise disruption.

The availability of trenchless techniques also means that other utilities can benefit from installing their services, particularly main distributors such as trunk cables installed in duct, at deeper levels to avoid the congestion of services found in the first 1.0 to 2.0 m below ground level.

A number of steps need to be taken before the selection of the most appropriate form of trenchless technique:

- site survey
- ground investigation to determine the soil and groundwater conditions
- site investigation to determine the location of existing pipelines, other services and potential obstacles
- inspection to determine the condition of the existing pipeline (if applicable).

3 The planning framework

Planning for the installation, replacement and renovation of underground services requires:

- establishing system/network performance criteria
- establishing system design criteria
- carrying out topographical surveys
- selecting preliminary routes
- collecting information on existing buried pipes and cables
- carrying out site investigation, including ground investigation and utility surveys, consider construction methods
- finalising site investigation
- finalising route selection and locations of major project features
- carrying out condition and locational surveys of existing services to be rehabilitated.

Early consideration needs to be given to the information required to procure and construct the work, much of which is also required during the planning stage. This includes the following:

- client requirements
- risk assessment (conceptual)
- contract terms
- a full ground investigation report
- statutory/by-law requirements
- noise restrictions
- settlement restrictions (ground, adjacent services and/or structures)
- reinstatement performance criteria (especially for deep trenches)
- access requirements
- traffic management requirements
- land use restrictions
- check on 'buildability' by consulting specialist contractor(s)
- establishing realistic contract period
- establishing the level of budgetary contingency
- planning communication channels for each stage of the project
- planning for serving notices and obtaining permits
- budgeting for the appropriate level of design and supervisory inputs.

The recommended procedures for carrying out the various aspects of the planning and investigation for the use of TME techniques for the installation, replacement or renovation of underground services are set out as a series of decision trees or flow diagrams in Figures 1 to 7.

Figure 1 shows the overall planning framework, from establishing the need for investment in uprating an underground services system, to the start of the detailed design and procurement stages. Figure 2 shows a flow diagram for the initial choice between installation, renovation or replacement when faced with a system rehabilitation task, and Figure 3 illustrates the choice between renovation or replacement. Figures 4 and 5 are frameworks for the planning and investigation processes to be used when considering new installation and replacement, and renovation, respectively.

Figures 6 and 7 are checklists providing guidance on the factors to be considered when selecting the most suitable method of installation, replacement or renovation for a particular project. When further detail is required, reference is made to the appropriate sections of these guidelines.

```
┌─────────────────────┐                              ┌─────────────────────┐
│ Establish required  │                              │  Establish existing │
│ system performance  │                              │  system performance │
│ and design criteria │                              │                     │
└─────────────────────┘                              └─────────────────────┘
           │                                                     │
           ▼                                                     │
┌─────────────────────┐                                         │
│ Evaluate against    │◄────────────────────────────────────────┘
│ existing system     │
│ performance         │
└─────────────────────┘
           │
           ▼
┌─────────────────────┐                              ┌─────────────────────┐
│ Identify shortfalls │─────────────────────────────►│ If no shortfall     │
│ requiring investment│                              │ continue to monitor │
│                     │                              │ performance         │
└─────────────────────┘                              └─────────────────────┘
           │
           ▼
┌─────────────────────┐
│ Develop preliminary │
│ solution(s)         │
└─────────────────────┘
           │
           ▼
┌─────────────────────┐
│ Choose between      │──────────────────────────────────────────► See Figure 2
│ installation,       │
│ replacement or      │
│ renovation          │
└─────────────────────┘
           │
           │              ┌─────────────────────┐
           │              │ Choose between      │──────────────────► See Figure 3
           │              │ replacement or      │
           │              │ renovation          │
           │              └─────────────────────┘
           │                        │
           ▼                        ▼                        ▼
┌──────────────┐        ┌──────────────┐        ┌──────────────┐
│ INSTALLATION │        │ REPLACEMENT  │        │ RENOVATION   │
└──────────────┘        └──────────────┘        └──────────────┘
           │                        │                        │
           ▼                        ▼                        ▼
        ┌─────────────────────┐              ┌─────────────────────┐
See Fig 4│ Planning and        │              │ Planning and        │ See Figure 5
◄────────│ investigation for   │              │ investigation for   │────────►
         │ 'new' system        │              │ renovating system   │
         └─────────────────────┘              └─────────────────────┘
                    │                                   │
                    ▼                                   ▼
         ┌─────────────────────┐              ┌─────────────────────┐
         │ Check against       │              │ Check against       │
         │ original criteria   │              │ original criteria   │
         └─────────────────────┘              └─────────────────────┘
                    │                                   │
                    ▼                                   ▼
         ┌─────────────────────┐              ┌─────────────────────┐
         │ Selected method,    │              │ Selected method and │
         │ route and principal │              │ principal details   │
         │ details             │              │                     │
         └─────────────────────┘              └─────────────────────┘
                    │                                   │
                    ▼                                   ▼
         ┌─────────────────────┐              ┌─────────────────────┐
         │ Risk assessment     │              │ Risk assessment     │
         └─────────────────────┘              └─────────────────────┘
                    │                                   │
                    ▼                                   ▼
         ┌─────────────────────┐              ┌─────────────────────┐
         │ Detailed design     │              │ Detailed design     │
         │ and procurement     │              │ and procurement     │
         └─────────────────────┘              └─────────────────────┘
```

Reassess if necessary   Reassess if necessary

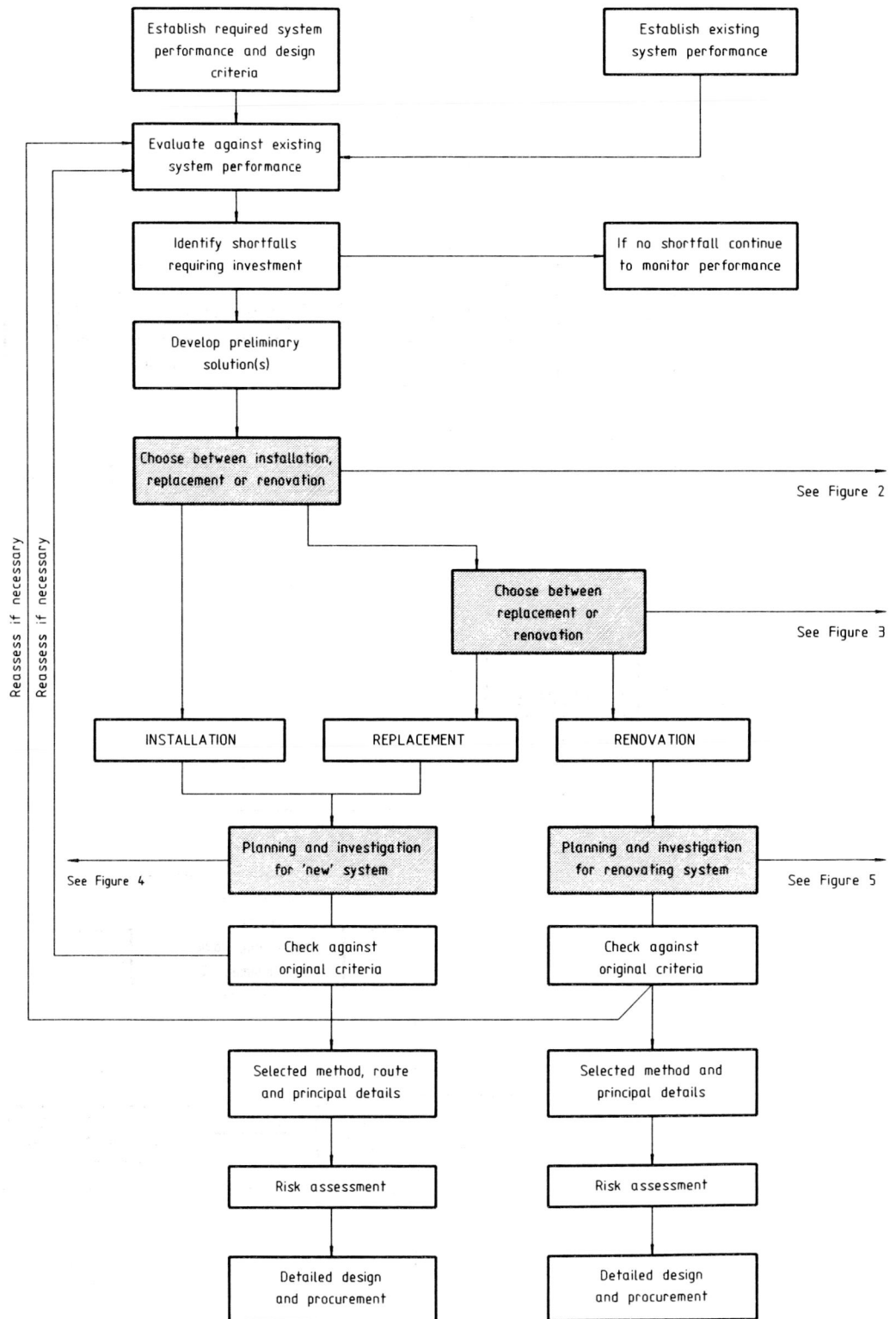

Figure 1 *Overall planning framework*

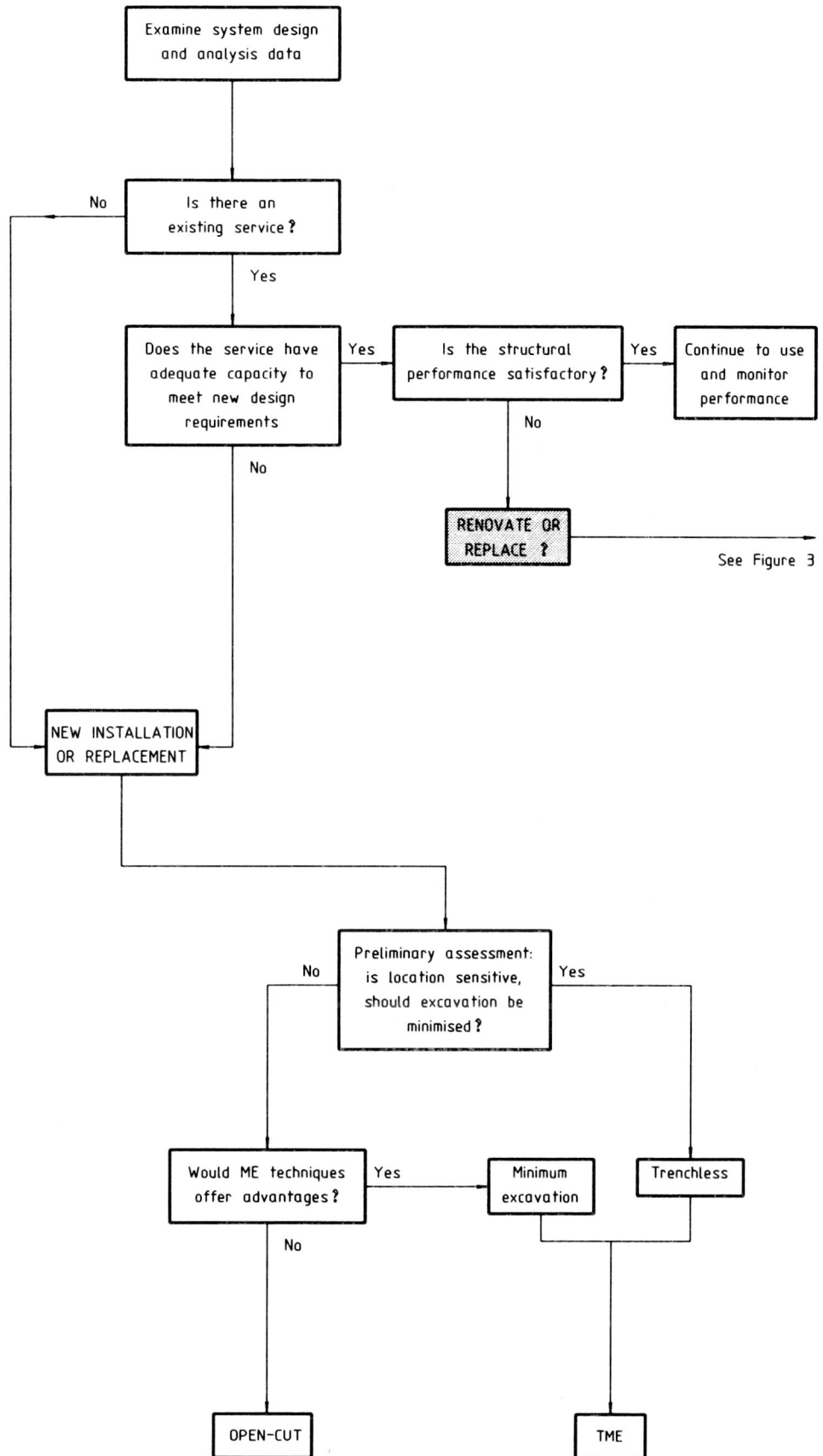

Figure 2 *Choice of installation, replacement or renovation*

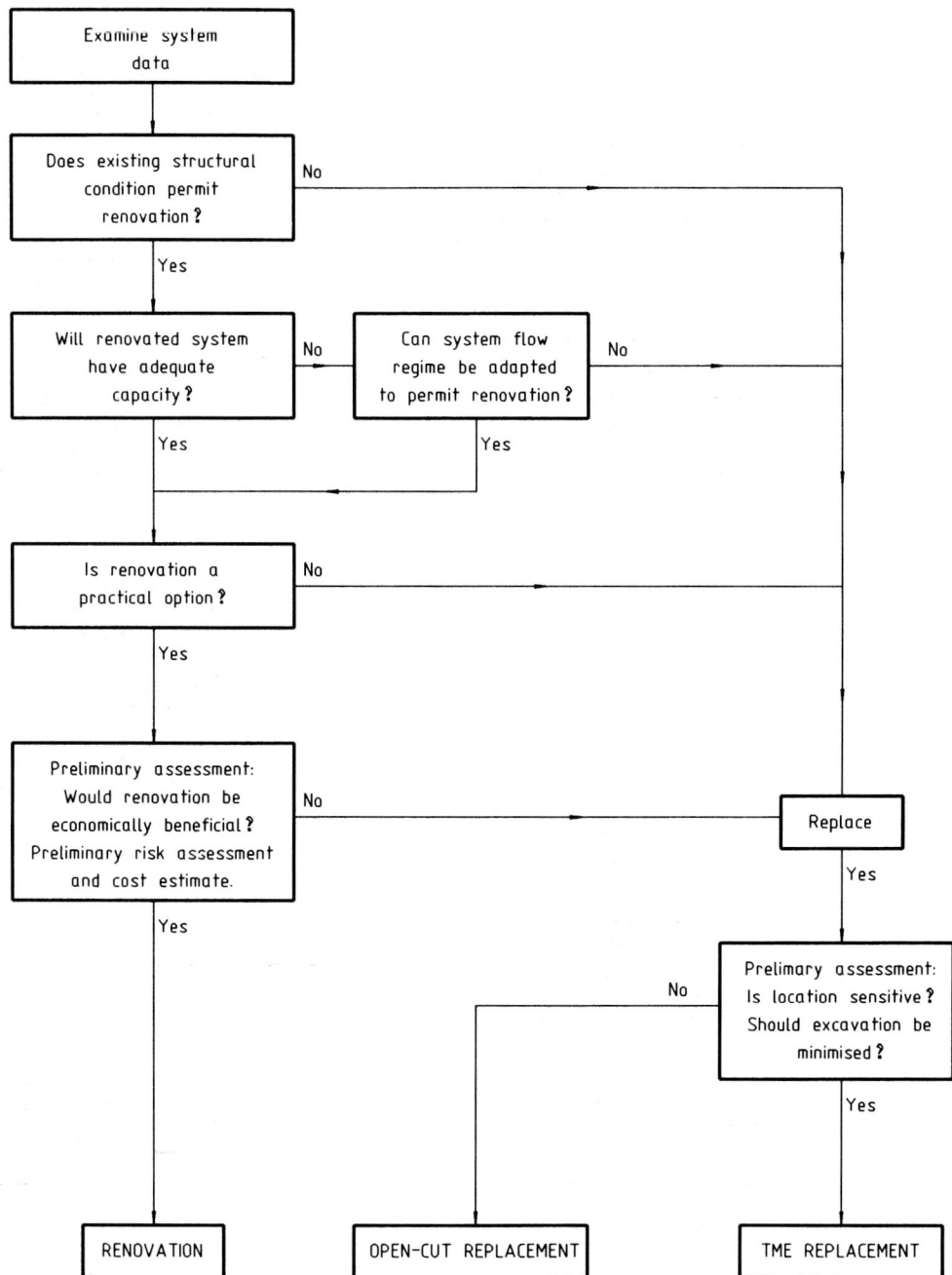

Figure 3 *Choice of renovation or replacement*

Establish type, size, depth
and length of service
to be installed. Preliminary
selection of suitable methods.

Collect details of restrictions
and requirements which will
apply throughout the project
See Section 3

Preliminary site survey
See Appendix B (B.2.1)

Preliminary locational survey
See Appendix B (B.2.3)

Preliminary selection of
routes and working areas

Risk assessment

Preliminary ground investigation
See Appendix B (B.2.2)

See Figure 6

Consider and select
suitable method(s)

Preliminary consideration of
social costs, if appropriate
use Figure 8

Reassess if necessary

Site survey
See Appendix B (B.2.1)

Locational survey
See Appendix B (B.2.3)

Ground investigation, to suit
proposed method(s)
See Appendix B (B.1.1 & B.2.2)

Check against original
criteria / assumptions

Finalise routes and
outline details of
selected method(s)

Cost method(s)

Consider social costs,
if appropriate use
Figure 8

Select method(s)

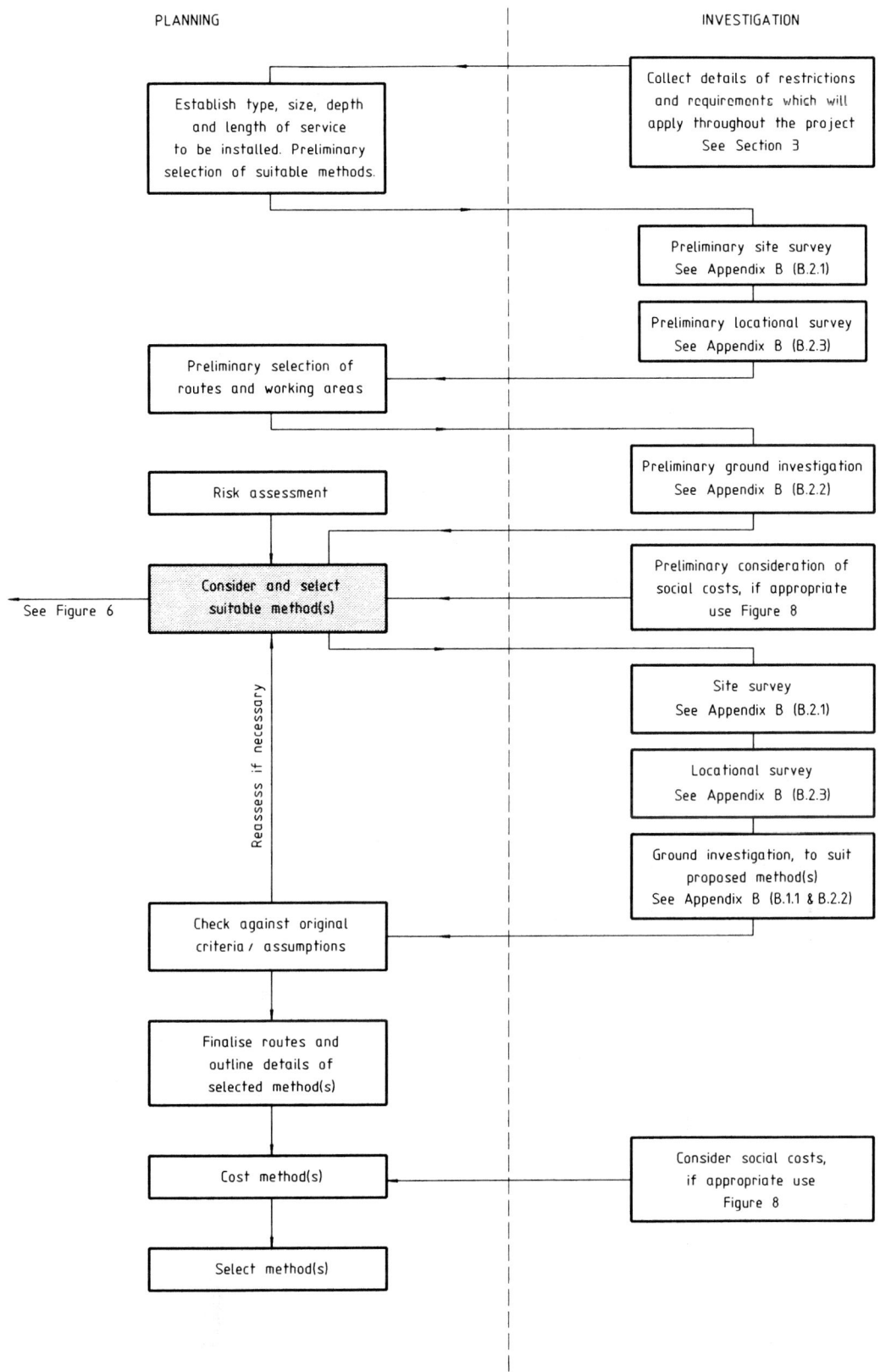

Figure 4 *Planning and investigation for new installation and replacement*

```
                                                    ┌─────────────────────────────┐
┌─────────────────────────────┐                     │ Collect details of restrictions │
│ Establish type, size, depth │                     │ and requirements which will  │
│ and length of service       │                     │ apply throughout the project │
│ to be renovated. Preliminary│                     │ See Section 3                │
│ selection of suitable methods.│                   └─────────────────────────────┘
└─────────────────────────────┘

                                                    ┌─────────────────────────────┐
                                                    │        Site survey          │
                                                    │    See Appendix B(B.2.1)    │
                                                    └─────────────────────────────┘

                                                    ┌─────────────────────────────┐
                                                    │      Locational survey      │
                                                    │    See Appendix B(B.2.3)    │
                                                    └─────────────────────────────┘

                                                    ┌─────────────────────────────┐
                                                    │      Condition survey       │
                                                    │    See Appendix B(B.2.4)    │
                                                    └─────────────────────────────┘

                                                    ┌─────────────────────────────┐
┌─────────────────────────────┐                     │ Preliminary ground investigation, │
│       Risk assessment       │                     │       if appropriate.       │
└─────────────────────────────┘                     │    See Appendix B(B.2.2)    │
                                                    └─────────────────────────────┘

                  ┌─────────────────────────────┐   ┌─────────────────────────────┐
See Figure 7      │   Consider and select       │   │  Preliminary consideration  │
                  │     suitable method(s)      │   │       of social cost.       │
                  └─────────────────────────────┘   │  If appropriate see Figure 8 │
                                                    └─────────────────────────────┘

                                                    ┌─────────────────────────────┐
                                                    │     Ground investigation,   │
                                                    │       if appropriate        │
                                                    │    See Appendix B(B.2.2)    │
                                                    └─────────────────────────────┘

┌─────────────────────────────┐
│      Finalise outline       │
│       details of            │
│     selected method(s)      │
└─────────────────────────────┘

┌─────────────────────────────┐                     ┌─────────────────────────────┐
│        Cost method(s)       │                     │    Consider social costs,   │
└─────────────────────────────┘                     │    if appropriate use       │
                                                    │          Figure 8           │
                                                    └─────────────────────────────┘
┌─────────────────────────────┐
│       Select method(s)      │
└─────────────────────────────┘
```

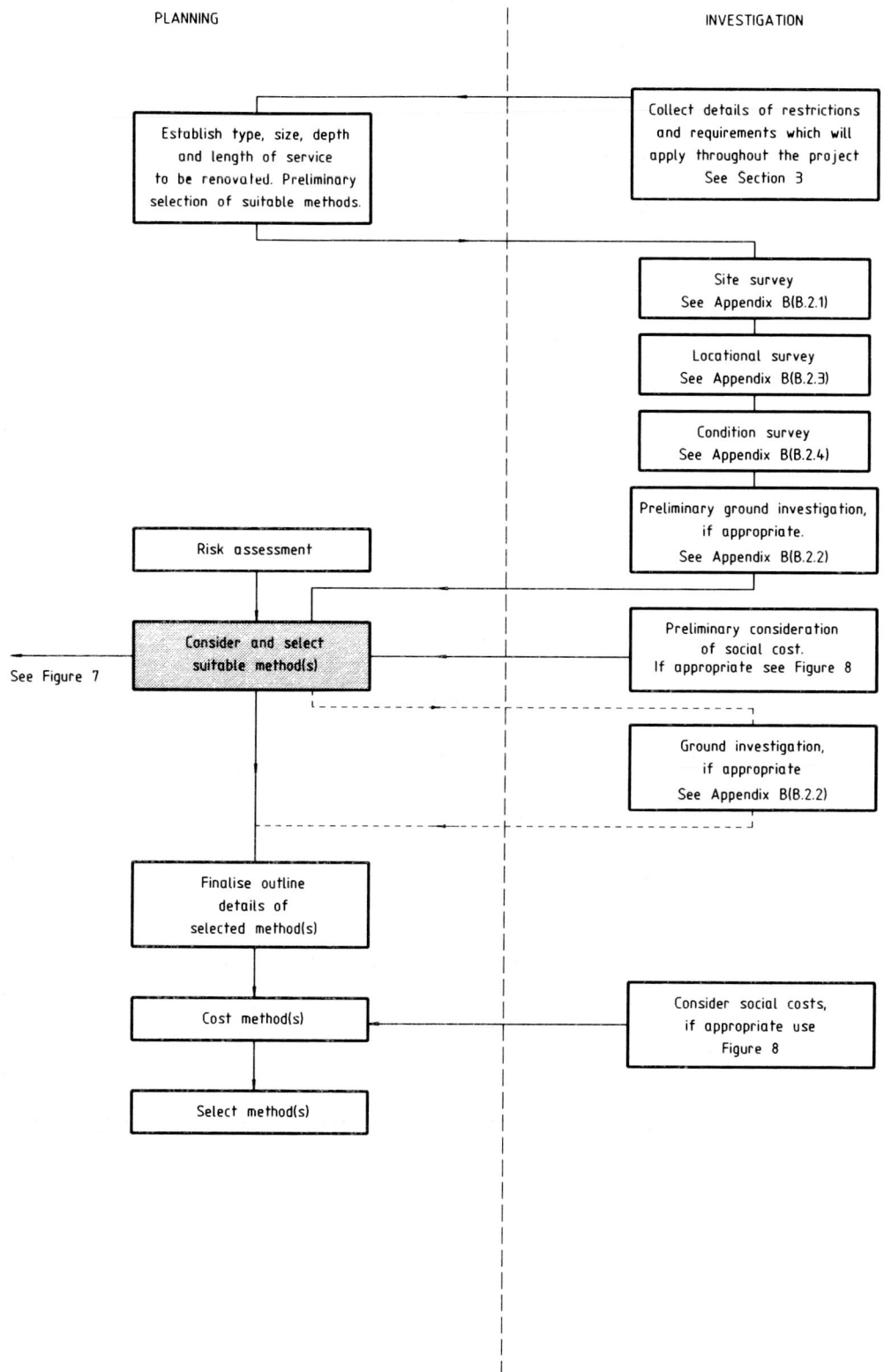

Figure 5 *Planning and investigation for renovation*

FACTORS TO BE CONSIDERED	REFERENCE IN GUIDELINES
LEGISLATION	
• The New Roads and Street Works Act 1991: 'Traffic sensitive'	
• Health and safety aspects	
• Environmental assessment	
TECHNICAL	
• Type of service - gravity pipeline - pressure pipeline - duct	Section 2.1
• Physical factors - size - length - depth - materials of construction	Appendix A: Table 1 Table 2 Table 3
• Accuracy	
• Size of temporary working areas	
• Resources (power, water etc)	
• Ground conditions	
DIRECT COSTS	Appendix D
SOCIAL AND ENVIRONMENTAL CONSIDERATIONS	Appendix D: Figure 8
ASSESSMENT OF RISK	Appendix E

Figure 6 *Checklist for selection of methods for new installation and replacement*

FACTORS TO BE CONSIDERED	REFERENCE IN GUIDELINES
LEGISLATION	
• The New Roads and Street Works Act 1991: 'Traffic sensitive'	
• Health and safety aspects	
• Environmental assessment	
TECHNICAL	
• Type of service - gravity pipeline - pressure pipeline - duct	Section 2.1
• Physical factors/condition - size, length, depth - connections - flows/pressures to be maintained - materials of construction	Appendix A: Table 1 Table 2 Table 3
• Type of fault to be repaired - continuous - local - serviceability - structural	
• Ground conditions (if applicable)	
• Size of temporary working areas	
DIRECT COSTS	Appendix D
SOCIAL AND ENVIRONMENTAL CONSIDERATIONS	Appendix D: Figure 8
ASSESSMENT OF RISK	Appendix E

Figure 7 *Checklist for selection of methods for renovation*

Appendix A Trenchless and minimum excavation techniques

Brief details of the various methods available are discussed below. Greater detail can be found in the *Introduction to Trenchless Technology*[2] and CIRIA Technical Note 127 *Trenchless construction for underground services*[3].

Quoted figures for system performance are those that were generally achievable in 1993. However, because of the rapid development of many of the techniques described they should be considered as being indicative only.

A.1 NEW INSTALLATION

Ground conditions are of prime importance when considering the suitability of various TME techniques. Table 2[4] shows the suitability, or otherwise, of a number of techniques for a range of ground conditions.

Table 1 *Suitability of techniques to various ground conditions*

Ground Conditions	Microtunnelling	Auger boring	Impact moling	Thurst boring	Directional drilling	Narrow trenching
Soft to very soft clays, silts and organic deposits	◐	◐	●	◐	◐	●
Medium to very stiff clays and silts	○	○	○	○	○	○
Hard clays and highly weathered shales	○	○	●	◐	○	○
Very loose to loose sands above and below the water table	◐	●	●	○	◐	●
Medium to dense sands above the water table	○	○	◐	○	○	●
Medium to dense sands below the water table	○	●	◐	○	○	●
Gravels and cobbles < 50 - 150mm dia	○	○	◐	◐	○	●
Soils with significant cobbles and boulders > 100 - 150mm dia	◐	◐	●	●	◐	●
Weathered rocks, marls, chalks and firmly cemented soils	○	○	●	●	◐	◐
Slightly weathered to unweathered rocks	◐	◐	●	●	●	◐

○ GENERALLY SUITABLE: by experienced contractor with suitable equipment.

◐ DIFFICULTIES MAY OCCUR: some modifications of equipment and/or operating procedures may be needed.

● SUBSTANTIAL PROBLEMS: generally unsuitable or not intended for this application.

Microtunnelling

Remotely controlled mechanical tunnelling systems for installing pipes up to 1000 mm-diameter. The excavated spoil is removed from the face by an auger or a pumped slurry system. The new pipeline is advanced by pipe jacking.

There is a wide range of systems available on the market capable of installing pipes from 150 mm diameter to well over the 1,000 mm-diameter limit set for this study. Most of the common pipe materials (concrete, clay, GRP, ductile iron and steel) can be installed by microtunnelling by using the appropriate pipe and jointing systems. Drive lengths for auger systems are generally 80–100 m, although lengths up to 120 m could be considered in good ground conditions. For pumped slurry systems, maximum drive lengths can be up to 120 m for the smaller diameters, whereas above 600 mm-diameter drive lengths in excess of 180 m have been achieved.

Approximate working and reception shaft sizes, and working compound sizes, are set out in Table 2.

Table 2 *Microtunnelling: typical drive and reception shafts, and working compound sizes*

Pipe dia. nominal	Drive shaft min dia.	Reception shaft min dia.	Minimum site layouts	
(mm)	(m)	(m)	Open ground (m × m)	Min width in roads (m)
250/300	2.75	2.5	15 × 10	4
450/500	3.0	2.75	15 × 10	5
600/675	3.0	2.75	20 × 10	5
900	4.3	3.0	30 × 10	5
975/1000	4.3	3.0	30 × 10	5

Auger boring

A boring machine containing a rotating cutting head is used to excavate the soil, which is transported back to the drive pit by auger flights operating in a casing. Pipes of 100–1000 mm diameter can be installed by pipe jacking.

The system is essentially unsteered and is generally only considered for short drives, although drives up to 75–100 m have been achieved.

Launch pits of 2–5 m length (depending on pipe size and type of boring machine and jacks) and of 2 m width are required. A smaller reception pit (2–4 m long) is used to recover the cutting head. Working compounds of approximately 75 m^2 are required.

Impact moling

The earth displacement (impact) mole is a cylindrical percussive hammer driven through the ground by compressed air to form a bore in which can be placed various kinds of supply conduits. Moles of sizes 30–80 mm diameter have been developed, which can also achieve 200–250 mm diameter by making multiple passes. The great majority of impact moles are unsteered.

The system is used by all utilities for short drives of up to approximately 40 m in length to make road crossings and is used for service house connections to avoid disturbance to drives and gardens.

Launch pits are typically 2 m long but this depends upon the pipe string to be installed (lead-in trenches may be required) and the length of the mole.

Rod pushing

A rod is advanced by a straight hydraulic push to form a 50 mm-diameter pilot hole, which is then back-reamed out to the required size.

The system is unsteered and is used to install pipes and conduits up to 150 mm diameter over lengths of 30–40 m.

Pipe ramming and thrust boring

A casing, usually steel, is driven through the ground by a pneumatic hammer (pipe ramming) or a straight hydraulic push (thrust boring). For smaller diameters the end of the casing may be closed, but in the larger sizes spoil is removed from the open-ended casing by compressed air and water, or an auger system.

The systems are unsteered and are used to install steel pipes between 50–1000+ mm diameter in which service pipes or cables are subsequently installed.

Launch pits vary between 3 m × 1 m and 9 m × 2 m, depending upon the system being employed. Reception pits can be much smaller because no excavation equipment has to be received.

Horizontal directional drilling

A rotating and steerable drill bit is launched at an angle of 10–15° from the surface to drill a 90 mm-diameter mud-filled pilot hole. A 125 mm-diameter washover pipe is drilled over the pilot string and following some 100 m behind the head. Alternate drilling continues first horizontally and then at an uphill angle until the pilot and the washover pipe have reached the exit point. The bore is enlarged by a rotating barrel reamer, with drilling mud being used to support the reamed hole. The service pipe is subsequently pulled into place through the bore.

This specialist technique is used primarily to install pressure pipelines under major obstacles such as motorway intersections, large rivers and airport runways. Lengths of 1–2 km have been successfully achieved.

Guided drilling

Small-diameter jets, mechanised cutting tools or displacement heads attached to a flexible drill string are positioned to form a bore as the head is thrust forward. The drilling head is launched from the surface at an inclined angle. Steering, in both horizontal and vertical planes, is achieved by controlling the rotation of the eccentric face of the head. Monitoring of the alignment is by a hand-held detector on the surface above the head. Back-reaming equipment is drawn through the pilot hole so that the service pipe, duct or cable can be installed.

The system can be used to install services between 40 mm and 150 mm in diameter.

Narrow trenching

A continuous chain or rock wheel trenching machine is used to excavate a narrow slot trench in both roads and fields. A range of machines is available capable of excavating trenches between 75 mm and 800 mm wide to adjustable depths and profiles up to a maximum of 2 m deep. This type of trenching system is only suitable for stable ground where trench support systems are not required for stability (support may be required for system). The system is most commonly used for water mains, gas mains, sewage pumping mains for non-gravity services and field drains to install

MDPE, PVC and ductile iron pipes. A space of approximately 50 mm is required either side of the pipe barrel to accommodate the pipe bedding material, usually sand or pea gravel. It is preferable to excavate, pipelay and backfill as quickly as possible; this is often carried out in a continuous train operation.

This type of trenching system can be very fast and excavation rates of 400 m per day have been achieved depending on whether in road, verge or field, and on the presence of other services. Working widths of 5–7 m are generally required in fields. In roads the minimum width is governed by the machine width (1.65 m) and safety considerations, typically 2.5–3.0 m.

A summary of the various new installation techniques discussed above is included as Table 3.

A.2 ON-LINE REPLACEMENT

In urban areas congestion from existing services and chambers and the high cost of trench reinstatement often makes it uneconomic to replace a defective or inadequate service on a new line. The existing hole in the ground therefore becomes a valuable asset, by offering a route for replacement as 'size-for-size' or 'up-sizing'. Any side-connections must be disconnected in advance and subsequently remade or rerouted.

Pipe bursting

An expanding device (pneumatic or hydraulic) is introduced into the defective pipeline, shattering the pipe and drawing in the new pipe. The potential damage to other services particularly due to the vibration of a pneumatic tool is a cause for concern. The technique is only suitable for certain pipes, e.g. cast iron, clay and concrete. It is not suitable for ductile iron or plastics. Rubber jointing rings and steel collars can also cause problems, therefore if a pipe has had several repairs it may not be suitable for pipe bursting.

Pipe eating

In this on-line microtunnelling replacement system the existing defective or inadequate pipe is crushed and removed through the new pipeline. This system is specifically designed for up-sizing.

A.3 RENOVATION

Where the performance of a pipeline is unsatisfactory but the fabric has a residual value either structurally or as a lining support, then renovation may be considered. A number of techniques are available.

Sliplining

A new pipeline of small diameter is inserted into the defective pipe and the annulus between the new and existing pipe grouted if required. The system has the merit of simplicity and is relatively inexpensive but there can be a significant loss of cross-sectional area, which can affect the hydraulic capacity, because the maximum outside diameter of the new pipe is determined by the smallest diameter of the existing defective pipe.

Table 3 Summary of trenchless construction techniques for new installations

Method	Ground conditions (see Table 2)	Typical length of drive	Steering and accuracy	Lining size and type	Access pit requirements	Comments
1. Microtunnelling: slurry-type tunnelling machine	Soft material at considerable depths below groundwater levels; some machines can cope with harder material and cobbles	100–120	Steerable; ±30 mm	250–1000+ mm dia jacked pipes	Variable; see Table 3	A variety of machines available with differing capabilities
2. Microtunnelling: auger-type tunnelling machine	Mainly soft material; not suitable for use more than 1.5 m below groundwater level; some machines can cope with soft rock and cobbles	80–100	Steerable; ±30 mm	150–1000+ mm dia jacked pipes	Variable; see Table 3	A variety of machines available with differing capabilities
Auger boring	Most soils	75–100	Not steerable; about 1.5% to 2% of drive length	100 mm dia upwards; normally jacked steel pipe	Variable; typically 5.0 × 2.0 m	An established method, but can be inaccurate
Impact moling	Most clays, silts, sands and gravels; isolated cobbles and boulders may deflect mole	Up to 70; normally about 40	Not steerable; within 1% of bore	30–250 mm PE pipe	Variable; typically about 2.0 m long	Extensively used by utilities for small-diameter mains and services for crossing roads
Rod pushing	Most soils	30–40	Not steerable	50–150 mm pipe	Variable; typically about 2.0 m long	Used for small-diameter pipes and conduits
Pipe ramming/thrust boring	Most soils	Up to 65; normally about 30	Not steerable	50–1,000+ mm steel pipe	Variable; 3.0 × 1.0 m min	Use limited to short crossings beneath railways and roads
Guided drilling	Most soils	Up to 150; normally about 60	Steerable	40–150 mm; normally plastic	None required	Used for small-diameter mains and services
Directional drilling	Most soils	250–1400	Steerable	50–1,000+ mm dia; steel or PE pipe	None required	Particularly suited for long crossings under rivers or canals
Narrow trenching	Soils not requiring stability support system	N/A	N/A	50–1000+ mm dia: MDPE, PVC or DI	Fields: 7 m Roads: 2.0–3.0 m width	Particularly suited for long lengths in areas uncongested by existing services

To some extent, the reduction in cross-sectional area can be offset by the lower frictional resistance of the liner. The choice of sliplining technique can have an impact on social costs. Most techniques require a lead-in trench and continuous strings of pipe are laid out along the road, which can cause access problems. However, some systems use short sections of pipe that can be inserted from existing manholes, thereby causing less surface disruption.

Modified sliplining

A family of techniques, including rolldown, swagelining and deformed lining, that uses the properties of PE or PVC to allow a temporary reduction in diameter or change in shape before insertion in the defective pipe. The inserted pipe is subsequently expanded to form a tight fit against the wall of the original pipe, thereby avoiding the need for annulus grouting. These systems are generally only suitable where the existing line is of good shape. Avoidance of the need for annulus grouting also means that the loss of hydraulic capacity is reduced when compared with conventional sliplining.

Lining formed in place

A resin-impregnated sock is inserted into the pipe or conduit (most shapes can be accommodated) and subsequently forced against the wall using air or water pressure to provide a continuous film lining. The resin is then cured and lateral connections reopened remotely. The method can accommodate large-radius bends and is used from 80 mm diameter upwards, principally on sewers, drains and gas mains. For water mains, only some systems currently have approval in some countries.

Where the integrity of an existing pressure pipeline can be relied upon and it is only necessary to prevent leakage, a loose-fit reinforced hose may be used. This is pulled through the existing line in a collapsed state and pressurised when in position. This system has application for pipelines in the range of 100–355 mm diameter.

A system using spirally wound plastic lining has been developed for defective sewers and drains. This uses a continuous PVC profile strip introduced by a helical winding machine at the bottom of a manhole into a pipeline. It has the advantages of not requiring special road openings and of being able to accommodate variations in cross-section and large-radius bends. It is usually necessary to grout the annulus and there is often significant loss of capacity. A variation of the system allows expansion, by reverse winding of an inserted undersized lining to form a close fit with the existing pipe, avoiding the need for annulus grouting. The system has been used in the range between 100 mm and 2500 mm diameter. Lengths of more than 100 m can be dealt with between access pits.

Spray-on lining

This is a technique for applying either an anti-corrosion or structural lining to pipes. A high-speed rotating head is winched through the pipeline, centrifugally depositing the lining material over the full internal surface. For more than 50 years, cement mortar has been used for lining water mains from 75 mm diameter upwards, to prevent encrustation, discoloured water and loss of capacity. More recently an epoxy resin material has been approved for use in potable water mains and is being used for sizes up to 600 mm diameter. Lengths of more than 100 m can be dealt with between access pits.

Structural spray-on lining is a new technique in sewer rehabilitation, stemming from advances in equipment and polymer formulation. The defective sewer receives a high-build spray lining of quick-setting epoxy resin or polyurethane material. The operation is carried out from the surface and from within existing manholes, thus avoiding road openings. It has been developed for use in the range of 150–600 mm diameter.

Localised repairs

Local defects may be found in an otherwise sound pipeline, due to cracking or joint failure. Systems are available for remote control resin injection to seal localised defects in the range 100–600 mm diameter. There are also sophisticated robotic systems for cutting grooves, filling with epoxy resins and smoothing the repair inside pipes from 150 mm diameter upwards.

Chemical grouting with urethane and similar materials has been used for many years in sewer rehabilitation, particularly in North America. Remote and man-entry grouting of defective joints and cracks may prevent infiltration and the formation of external voids in an otherwise sound pipeline.

Systems are also available which are a localised form of modified sliplining in which a short section of liner is sprung into place at the defective length and for a localised form of lining formed in place.

Chemical stabilisation

The stability of sewers in doubtful ground conditions may be restored by treating manhole lengths with a two-part chemical dosing. The sewer length is sealed and filled first with solution A, which is then pumped out and the length refilled with solution B, which is also subsequently pumped out. The chemical reaction between the components seals joints and cracks in the pipe and stabilises the surrounding soil.

Appendix B Investigation strategies and methods

B.1 INVESTIGATION STRATEGIES

B.1.1 New installation and replacement

The documents that guide the investigation strategies for new installation and replacement include:

- BS5930 (Code of practice for site investigations)[5]
- BS1377 (Methods of tests for soils etc)[6]
- New Roads and Street Works Act 1991
- British Gas Codes of Practice
- In-house guidance notes and specifications.

The investigation work is required as a minimum to determine:

- ground and groundwater conditions, including the possibility of rising groundwater levels
- the position of existing pipelines, services and other obstacles.

There are three main types of investigation commonly used for trunkless installation projects:

- site survey – topographic and hydrographic surveys
- locational survey – use of non-intrusive tracing and confirmatory trial holes
- ground investigations – trial pits, borings, and soil sampling and testing.

The investigation process is normally carried out in two stages, comprising a preliminary stage followed by the full-scale investigation. Preliminary-stage investigations will consist of desk studies and field visits. Full-scale investigations are usually carried out by a specialist contractor.

B.1.2 Renovation

The design procedures for the renovation of sewers are set out in the *Sewerage Rehabilitation Manual*[7], and in particular Chapter 2 of Volume III. The procedures follow the design process, from identified performance deficiencies, to the selection of the most appropriate renovation techniques. The procedure is not divided into preliminary and detailed design phases since it is considered essential that the planner and designer should understand the fundamental advantages or limitations of the renovation options at all stages of the scheme design.

Information to be obtained for the renovation of existing services by trenchless techniques usually includes:

- details of the level of serviceability that is being sought following renovation

- details of the length, size, position and depth of the pipeline and its access arrangements

- up-to-date details of the structural and service condition of the existing services, often obtained by CCTV and/or sonar surveys

- details of design conditions including flow rates and the composition and concentration of any special fluids or gases to be conveyed

- local ground conditions including groundwater levels if these are pertinent to the structural or serviceability design of the particular renovation system

- details of flows to be temporarily accommodated during installation of the renovation system to maintain the overall level of service in the utility system (e.g. by overpumping) including, if appropriate, response to varying rainfall intensities

- availability of mains water.

B.2 INVESTIGATION METHODS

B.2.1 Site surveys

The techniques and aids most commonly used to establish appropriate sites for access pits, drive and reception shafts, and working areas are:

- land ownership/occupation records

- current OS maps

- geographical information systems (GIS)

- historical maps and records

- aerial photography

- site visits and topographical surveys

- details of current and historical usage

- discussions with local people

- reference to other bodies e.g. EA and local regulatory authorities.

These techniques are common to open-cut trenching and most other forms of civil engineering construction.

B.2.2 Ground investigation

To be fully effective and reliable, all ground investigation work needs to be carried out by skilled, competent and experienced staff at all stages of:

- the initial planning and selection of the most appropriate form of investigation

- the execution of the on-site investigation

- the supervision of the on-site work (rapid and accurate data evaluation can establish the need for additional exploration and testing)

- the laboratory testing

- the preparation of factual reports

- the preparation of interpretative and engineering reports

- the provision of advice during the planning and design phases

- the continuing provision of advice and support during the construction phase.

The steps commonly involved in carrying out ground investigations are studies of:

- geological survey maps and memoirs

- existing borehole and well records held by the British Geological Survey

- in-house records

followed by:

- site examination of local geological features and any outcrops

- physical investigation.

The methods of physical investigation can be divided into the more traditional intrusive methods, in which the ground is disturbed, and the newer non-intrusive (geophysical) techniques. Intrusive methods available to determine the engineering characteristics of sub-surface soils are described in detail in BS5930 and other general texts.

In addition, the following non-intrusive techniques are sometimes used:

- seismic refraction

- ground-probing radar

- ultrasonic detection

- down-hole CCTV (for very specific applications)

- resistivity

- thermal imaging.

Non-intrusive techniques, their application and interpretation are based around highly specialised systems. Therefore, specialist advice needs to be sought on their suitability and accuracy for each individual use. In most circumstances these techniques should be used to complement traditional investigation techniques to give a full picture of conditions below the ground surface.

The Pipe Jacking Association (which represents both pipe jacking and microtunnelling interests in the UK)[8] recommends that a full site investigation be undertaken to determine the characteristics of the soils likely to be encountered, together with details of the water table, any tidal or seasonal changes, and the permeability of the soil. The installation of piezometers would be necessary to enable continued monitoring of pore water pressure changes. For particular types of soils it is recommended that ground material characteristics should include those detailed in Box 2.

Box 2 *Ground material characteristics to be determined for microtunnelling*[8]

Non-cohesive soils

Grading analysis of particle distribution
Permeability of soil
Soil density
Standard penetration test of soil
Predicted presence and size of cobbles and boulders
Pumping tests

Cohesive soils

Apparent cohesion or unconfined compressive strength
Soil density
Standard penetration test of soil
Mositure content
Plasticity indices
Predicted presence and size of cobbles and boulders

Mixed soils

Information as above for cohesive soils together with evidence of artesian or perched water tables and pumping tests.

Fill material

Information as above for cohesive soils with particular reference to compaction and the nature of the material, toxic constituents and the presence of gases.

Rock

Colour
Grain size
Geological type
Rock strength (MN/m^2)
Abrasivity
Total core recovery (TCR)
Solid core recovery (SCR)
Rock quality description (RQD)
Fracture index (FI)

B.2.3 Locational surveys

The techniques and aids most commonly used to locate existing services, pipes and other sub-surface obstructions are:

- utility records

- current and historical maps

- trial holes

- visual inspection for above-ground evidence of buried services including apparatus, covers, marker posts and reinstatement scars

- electromagnetic detection using hand-held scanners

- ground-probing radar (less common; more specialised).

The determination of the position of existing plant in preparation for trenchless work has much in common with that for open-cut work. There are, however, some differences:

- where renovation or on-line replacement is proposed, more detail is required about the condition and position of bends and fittings, etc.

- the progress of trenchless construction does not permit visual observation of the workface, so an uncharted obstruction may be more difficult to avoid

- trenchless installation methods generally require launch and reception pits and these may occupy a greater width than a trench, affecting locally more services than an equivalent trench

- the effects of heave, displacement or ground loss on buried plant are not observable

- in general, trenchless construction provides fewer opportunities for improving records of other existing buried plant by direct observation than does trenching.

B.2.4 Condition surveys

The techniques and aids most commonly used to determine the condition and other information required for renovation of conduits are:

- locational surveys (see B.2.3 above)

- CCTV surveys

- sonar surveys, if overpumping of sewers or drains is not practicable

- flow surveys (including leakage and infiltration testing)

- infrared thermography (less common)

- trial pits

- removal of sample section of pipe for materials analysis.

B.2.5 Supplementary surveys

If social costs are to be given serious consideration, details of traffic surveys, environmental impact studies and customer surveys may be required.

Appendix C Design considerations

Design considerations are given for new installation and replacement by microtunnelling and narrow trenching, together with some general planning considerations for renovation. Although the discussion concentrates upon the water industry, other utilities face similar problems and their planners need to address similar design factors.

C.1 NEW INSTALLATION AND REPLACEMENT

The potential advantages of TME techniques – less disruption, less risk and less damage – can only be realised by planning a project around the techniques from the start. This means that it is essential to go through a fairly detailed iterative selection procedure at the same time. The construction of sewers usually represents the most serious impact upon the public highway.

The trenchless technique normally used for the construction of sewers of diameters less than 1000 mm is microtunnelling. Other utilities may use different techniques to suit their particular circumstances. The advantages of microtunnelling and other techniques over conventional open-cut trenching have been discussed earlier, but they are summarised here as a reminder to the decision-maker:

(a) Concentrating construction operations on discrete, planned working sites reduces disruption, and in some cases this may mean that streets can remain open to traffic that might otherwise have had to be closed.

(b) Safety is improved by concentrating the working areas, which means that safety and protection of both the operators and the general public can be more easily established and controlled.

(c) Damage to the highway structure is concentrated at working sites, thereby reducing reinstatement provisions.

(d) Noise-creating activities are concentrated at working sites, thereby reducing reinstatement provisions.

(e) Pipelines can be installed to at least as high a level of accuracy as open-cut work, but, more importantly, the reduced disturbance and the absence of a granular bed means that the installed accuracy is maintained.

(f) Surface and sub-surface ground movements are usually very small (dependent upon a number of factors), thereby reducing the risk of damages to adjacent structures and other services.

Recognising these advantages, Yorkshire Water, Decon Engineering and ARC Pipes entered into a two-year Joint Venture for Microtunnelling in 1987. This culminated in Yorkshire Water's *Code of Practice for Microtunnelling*[9]. The Code of Practice offers guidance on sewerage scheme design, but its advice on general matters related to the installation of underground services is generally applicable to all trenchless techniques for new construction. The *Civil Engineering Specification for the Water Industry*[10] provides information on materials and workmanship for new installation and sewer renovation.

Narrow trenching is a high-speed trenching system primarily used for non-gravity services, in which the operations of excavation, pipelaying, backfilling and reinstatement are organised in rapid sequence. It can involve longer lengths of trench being open for a much shorter period of time than would be the case for conventional open-cut trenching. Although the highway authority may place a restriction on the length of excavation open at any time, this may need to be increased by negotiation in order for the advantage of this type of system to be maximised.

In many circumstances detailed, long sections of pipelines are not necessary for installation by narrow trenching and are often inappropriate because constant gradients can result in excessive excavation. A general profile of the pipeline is required, together with details of minimum depth, minimum gradient and other specific constraints. Using these criteria, the pipeline can normally be laid to follow the general ground profile.

Taking into account modern techniques for surveying, cleaning and repairing pipelines, it could be worth considering a relaxation of some design criteria, e.g. straight lines between manholes. This could enable the narrow trenching technique to be used for other applications such as sewers.

C.2 RENOVATION

During the renovation process the size and configuration of all access points into the underground system – manhole and chamber covers, for example – should be brought up to current standards. This should ensure a safer and more efficient working environment during both the actual renovation process and for subsequent operation and maintenance procedures. If appropriate, the opportunity could be taken to relocate the access point to a position where it is less disruptive to traffic and safer to gain entry.

It is essential to identify the exact size and material of the pipeline to be renovated. A common problem is ascertaining the precise condition of the pipe before renovation. For sewers, CCTV surveys are frequently used. However, the surveys are frequently two to three years old, collected as part of the drainage area planning process. Significant deterioration may have occurred between the study and the start of a renovation contract. Consequently, it would be advantageous for the client to resurvey the service immediately before the contract in order to provide up-to-date information on the condition of the pipe to the tenderers. Alternatively, the client could elect to have the survey carried out by the contractor at the start of the renovation contract but with provision in the contract for investigation, revised design (if necessary) and procurement of materials.

Defects in water mains are often revealed by bursts and leaks. Evidence can be obtained about the size, material and condition of the pipeline during repairs. Trial holes may also be needed.

Appendix D Costs

D.1 DIRECT COSTS

The direct costs associated with projects for the installation, replacement and renovation of services include the following:

- planning, design and supervision

- direct civil engineering costs (payments to contractors and suppliers)

- reinstatement of highways

- diversion of other statutory undertakers' services

- traffic management, diversions and accommodation works

- compensation to property-owners

- compensation to commercial concerns for loss of profits.

The first five items are normally accounted for to their full value. The remaining items represent ex gratia payments to third parties affected by the works. It is a commonly held view that for most works the total individual losses incurred by third parties exceed the total compensations actually paid. Any differences can be classed as additional social costs.

A number of documents refer to the direct costs of installing new or replacement services. However, most organisations prefer to use historic cost data for estimating purposes. Early discussions with specialist contractors can also assist in this process.

It is not the purpose of these guidance notes to produce detailed costings for the various methods of installation and rehabilitation being discussed. Such cost data can become rapidly out of date and hamper the decision-making process. Many organisations are establishing their own price databases to cover this type of work, and for most circumstances specialist contractors will be able to provide assistance in preparing budgetary estimates.

Factors which affect the unit costs of installing new underground services by trenchless techniques are:

- length of drive (set-up costs are generally independent of length)

- diameter

- type of pipe

- ground conditions

- number of set-ups (angular changes and moves)

- the number and type of other services encountered (usually only at shaft locations)

- depth (not so critical as for trenching).

Factors which affect the unit costs of rehabilitating underground services by trenchless techniques are:

- location and accessibility

- length

- diameter

- type, condition (including shape) and material of existing pipe

- ground conditions (in some instances)

- extent of infiltration

- number of connections

- number of chambers or manholes to be replaced

- any overpumping or pressurisation required

- depth of pipe, particularly if lead-in trenches are required

- cleaning of pipe

- use of CCTV survey

- discount rates, when whole-life costs are considered.

D.2 INDIRECT COSTS

Indirect costs associated with utility systems fall into two broad categories:

- those associated with system performance and not meeting the appropriate level of service to customers

- those associated with system rehabilitation due to the disruptive nature of engineering works required to install, replace or renovate underground services.

The first category is outside the scope of these guidance notes.

The second category of indirect costs is commonly referred to as social and environmental costs. It is widely accepted that these costs are greater in an urban environment than in a rural environment. However, the social and environmental impact on rural life, and on the environment in general, may be significant and planners will increasingly need to consider these aspects. For works carried out in the urban environment they fall under the following general headings:

- disruption of the local economy through loss of trade and increased costs of operation

- congestion, delays and diversions to traffic, including to public transport vehicles, with associated increases in traffic accidents and disruption to the wider community than that directly affected by the work

- latent damage to other underground utilities, adjacent buildings and the highway formation

- environmental damage from the works themselves and from any resultant traffic congestion and diversion, including loss of amenity due to noise, vibration, visual intrusion, dust, dirt and smell.

Key factors to be considered if overall costs are to be minimised include:

- traffic flows – volume, type of vehicles (including public transport), and variation

- reduction in available road width as a result of the work

- location of trenches or shafts within the carriageway

- location of shafts within the road network

- timing of works

- possible diversion routes

- the potential for work to be carried out on other underground utilities at the same time

- assessment of loss to businesses as a result of the proposed work

- access to residential, industrial, commercial and other property

- sensitivity of the area to noise, dust, and other adverse environmental impacts

- methods for making connections between the main service and individual properties.

There is no statutory requirement in the United Kingdom to give consideration to social and environmental costs during the appraisal processes at the planning, design or construction stages; except where environmental assessment studies are required by the Town and Country Planning (Assessment of Environmental Effects) Regulations 1988.[11] However, it is generally becoming accepted that social and environmental costs need to be considered in appraisal systems for works directly affecting the general public. Vickridge, Read and others,[12, 13] have carried out a number of studies into the social and environmental costs and benefits of trenchless techniques for the installation and renovation of sewerage systems. In their latest paper[14] they propose a procedure for evaluating indirect costs with particular emphasis on congestion costs to establish a protocol for setting charges for road space occupation. This framework is included as Figure 8. Traffic disruption is selected as the social factor to be considered, because it is the easiest cost to evaluate in a rational and structured way. Further research is needed on this aspect, but at present it offers a fairly straightforward method of evaluating total costs during project appraisal. If road space rental charges were adopted in the future, they would probably be calculated on a similar basis.

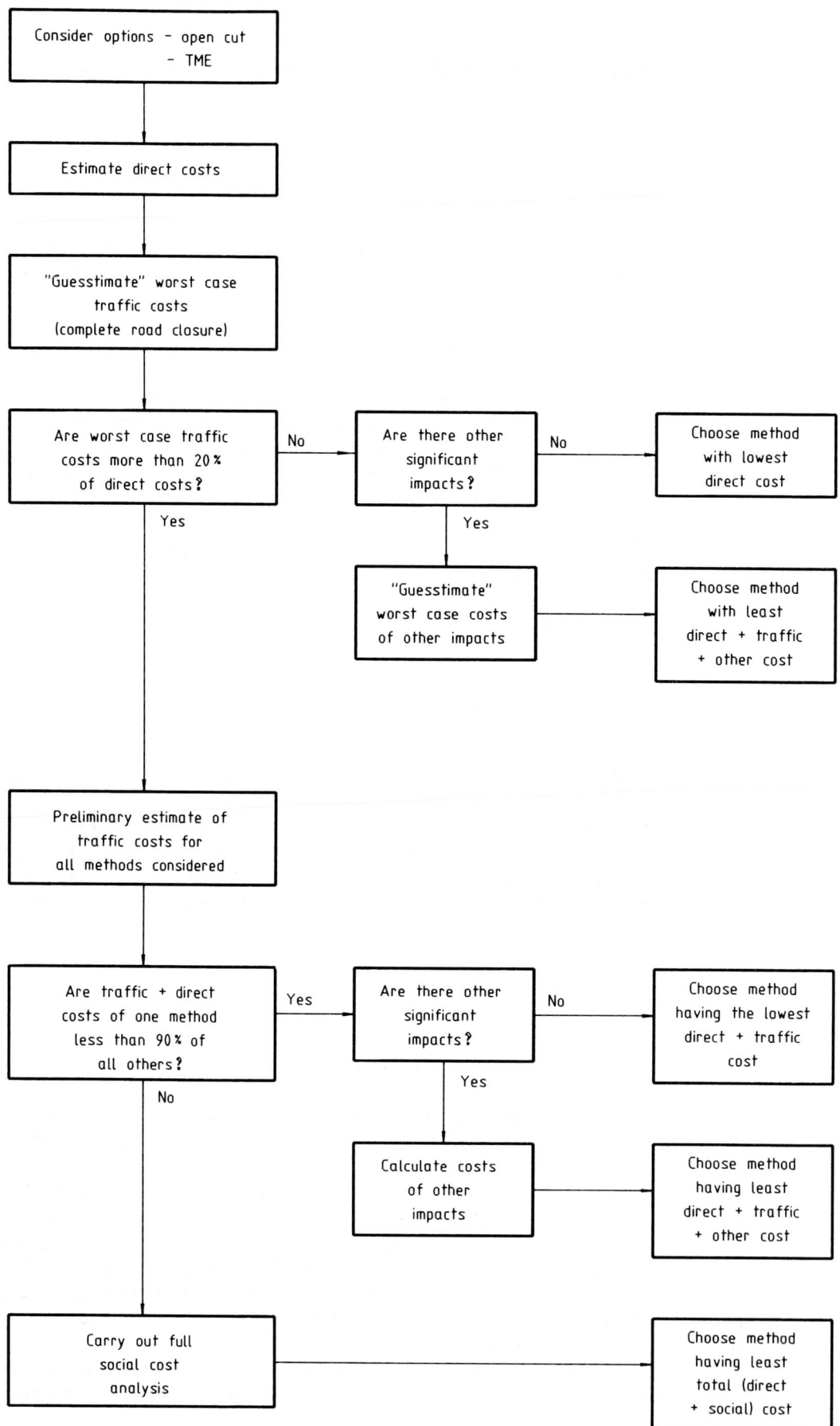

Figure 8 *Framework for inclusion of social costs in selection of method of construction*

Appendix E Risk assessment

Risk can be categorised as contractual or construction. Contractual risks can arise from inadequate contract preparation and administration. In broad terms, the risk increases with decreasing clarity of contract and can be dealt with through improving both contract clarity and administration practices. Construction risks arise from diverse factors such as weather, site conditions and construction methods. Those risks can often only be defined and apportioned, not eliminated.

The most cost-effective contract is one that allocates each risk to the party that is best able to manage and minimise the risk, recognising the unique circumstances of the project. In this regard, the general forms of contract commonly used for construction projects in the UK, e.g. the ICE *Conditions of Contract (Sixth Edition)*,[15] which make provision for the sharing of risk, can be applied to contracts using TME techniques.

The risks associated with installations of underground services by TME techniques may be summarised as follows:

Risk of failure

- due to mechanical failure of plant

- due to failure of materials – pipes or linings

- due to loss of directional control

- due to presence of material beyond the excavation or jacking capability of the system.

These risks can be mitigated by selecting the most suitable technique and materials of construction for each project. Quality assurance and control procedures should include selection of contractors (the right contractor with the right equipment), performance assessment of materials, contract strategy and appropriate levels of training of operatives and supervisory staff.

Risk of damage

- to other utility services

- to the road structure

- to adjacent structures and buildings

These risks can be mitigated by carrying out an appropriate level of site inspection and investigation to assess ground conditions and locations of other utilities and structures.

Most utility services, with the exception of main sewers and some other trunk services, are generally located no more than 1.0–2.0 m below road level. In most urban and developed areas in the UK this band of ground bears little relationship to the natural geological formation of the area. When small-diameter services are being installed, work is regularly undertaken without any site investigation. If a relatively short length of service is being installed, by impact moling for example, any installation problems can normally be dealt with by local excavation relatively cheaply. However, the same cannot be said if the moling causes damage to other utilities. Damage to fibre-optic cables and basement flooding caused by damaged sewers can be expensive to rectify, while damage to live gas and electricity services can prove dangerous and risk loss of life.

Recent insurance cases have demonstrated that although a client or contractor may believe that they are covered against third-party risks, insurers are taking a hard line especially when it is shown that no attempt was made at carrying out a site investigation. Consequently, a minimum requirement for an investigation for a minor trenchless installation should be the location of existing utilities or other underground property that may become damaged.

For the installation of major services using expensive systems, for example sewers by microtunnelling, preventing damage to third-party property is still important, but the higher risk is attached to potential equipment malfunction during installation. In these cases a full investigation to establish the ground conditions is of paramount importance.

References and selected bibliography

1.   *Roads and the utilities – review of the Public Utilities Streetworks Act 1950*
     Department of Transport, 1985

2.   INTERNATIONAL SOCIETY FOR TRENCHLESS TECHNOLOGY
     Introduction to trenchless technology, 2nd edition
     ISTT, 1992

3.   CONSTRUCTION INDUSTRY RESEARCH AND INFORMATION ASSOCIATION
     Trenchless construction for underground services
     Technical Note 127, CIRIA, 1987

4.   INGOLD, T S and THOMSON, J C (1989)
     Site investigation related to trenchless techniques
     In: *Proceedings of No-Dig 89 London*
     ISTT, Paper 3.1, 1989

5.   BRITISH STANDARDS INSTITUTION
     Code of practice for site investigations
     BS5930: 1981

6.   BRITISH STANDARDS INSTITUTION
     Methods of test for soils for civil engineering purposes
     BS1377: 1990

7.   WATER RESEARCH CENTRE and WATER AUTHORITIES ASSOCIATION
     Sewerage rehabilitation manual, 3rd edition
     WRc, 1994

8.   PIPE JACKING ASSOCIATION
     A guide to pipe jacking and microtunnelling design
     Pipe Jacking Association, 1987

9.   YORKSHIRE WATER
     Code of practice for microtunnelling, 2nd edition
     Yorkshire Water Plc, 1991

10. WATER AUTHORITIES ASSOCIATION
     Civil engineering specification for the water industry (CESWI), 4th edition
     WRc, 1993

11. Town and Country Planning (Assessment of Environmental Effects) Regulations 1988
     HMSO, SI1199, 1988

12. VICKRIDGE, I G, READ, G F, GREEN, C and WOOD, J (1987)
     Current research into the social costs of sewerage systems
     In: *Proceedings of No-Dig 87 London*
     ISTT, Paper 5.3, 1987

13. BRISTOW, A L, LETHERMAN, K M, LING, D J, READ, G F and VICKRIDGE, I G (1988)
   Social costs of sewerage rehabilitation – where can No-Dig techniques help?
   In: *Proceedings of No-Dig 88 Washington*
   ISTT, Paper 2C, 1988

14. VICKRIDGE, I G, LING, D J and READ, G F (1992)
   Evaluating the social costs and setting the charges for road space occupation
   In: *Conference Papers of No-Dig 92 Washington*
   ISTT, Paper B2, 1992

15. INSTITUTION OF CIVIL ENGINEERS, ASSOCIATION OF CONSULTING ENGINEERS, FEDERATION OF CIVIL ENGINEERING CONTRACTORS
   Conditions of contract and forms of tender, agreement and bond for use in conjunction with works of civil engineering construction, 6th edition
   Thomas Telford Ltd, 1991

The following sources provide general background to the planning and investigation process.

BRITISH STANDARDS INSTITUTION
Sewerage
BS8005: Parts 0, 1 and 2: 1987
BS8005: Part 3: 1989
BS8005: Part 5: 1990

BRITISH STANDARDS INSTITUTION
Code of practice for safety in tunnelling in the construction industry
BS6164: 1990

BRITISH STANDARDS INSTITUTION
Code of practice for pipelines (Pipelines on land: general)
BS8010: Part 1: 1989

Construction (Design and Management) Regulations 1994
HMSO, 1994

Control of Substances Hazardous to Health Regulations 1988
HMSO, SI1657, 1988

DEPARTMENT OF TRANSPORT, SCOTTISH OFFICE INDUSTRY DEPARTMENT, WELSH OFFICE AND DEPARTMENT OF THE ENVIRONMENT (NORTHERN IRELAND)
Specification for highway works
HMSO, 1991

Electricity Supply Regulations 1988
HMSO, SI1057, 1988

HEALTH AND SAFETY EXECUTIVE
Avoiding danger from underground services
Booklet: HS(G)47, HMSO

KRAMER, S R, McDONALD, W J and THOMSON, J C
An introduction to trenchless technology
Van Nostrand Reinhold, 1992

NATIONAL JOINT UTILITIES GROUP
Recommended positioning of utilities' mains and plant for new works
Publication No 7, NJUG

STEIN, MOLLERS, BIELECKI (1989)
Microtunnelling
Ernst & Sohn, 1989

THOMPSON, J C
Pipejacking and microtunnelling
Blackie Academic & Professional, 1993

WATER SERVICES ASSOCIATION, FOUNDATION FOR WATER RESEARCH and
WATER RESEARCH CENTRE
Water mains rehabilitation manual
Water Research Centre, 1989

WATER SERVICES ASSOCIATION
Sewers for adoption – A design and construction guide for developers, 4th edition
WRc, 1995